I0814538

Inside the Military

MILITARY ROBOTS

Emma Bassier

DiscoverRoo
An Imprint of Pop!
popbooksonline.com

abdobooks.com

Published by Pop!, a division of ABDO, PO Box 398166, Minneapolis, Minnesota 55439.

Printed in the United States of America, North Mankato, Minnesota.

052019
092019

Cover Photo: Shutterstock Images

Interior Photos: Shutterstock Images, 1; Defense Visual Information Distribution Service, 5, 6, 7, 14, 15, 20, 22, 25, 29, 31 (left); US Air Force, 9, 12–13, 16 (top), 30; US Army, 10; Stocktrek Images, Inc./Alamy, 11; Historic Collection/ Alamy, 16 (bottom); Thierry Falise/LightRocket/Getty Images, 17 (top), 26, 31 (right); Mike Derer/AP Images, 17 (bottom), 23; iStockphoto, 19; Bob Edme/AP Images, 21; Tactical Robotics/Cover Images/Newscom, 27; Ben Birchall/PA Wire URN:40157685/AP Images, 28

Editor: Connor Stratton
Series Designer: Jake Slavik

Library of Congress Control Number: 2018964856

Publisher's Cataloging-in-Publication Data

Names: Bassier, Emma, author.

Title: Military robots / by Emma Bassier.

Description: Minneapolis, Minnesota : Pop!, 2020 | Series: Inside the military | Includes online resources and index.

Identifiers: ISBN 9781532163852 (lib. bdg.) | ISBN 9781644940587 (pbk.) | ISBN 9781532165290 (ebook)

Subjects: LCSH: Military robots--Juvenile literature. | Robotic soldiers--Juvenile literature. | United States--Armed Forces--Robots--Juvenile literature.

Classification: DDC 623.04--dc23

WELCOME TO DiscoverRoo!

Pop open this book and you'll find QR codes loaded with information, so you can learn even more!

Scan this code* and others like it while you read, or visit the website below to make this book pop!

popbooksonline.com/military-robots

*Scanning QR codes requires a web-enabled smart device with a QR code reader app and a camera.

TABLE OF

CONTENTS

CHAPTER 1

MILITARY ROBOTS IN ACTION

A soldier presses a button on a remote. Nearby, a huge vehicle comes to life. The soldier uses the robot to dig through the dirt. Other robots clear away **mines**. Together, they create a safe path.

WATCH A VIDEO HERE!

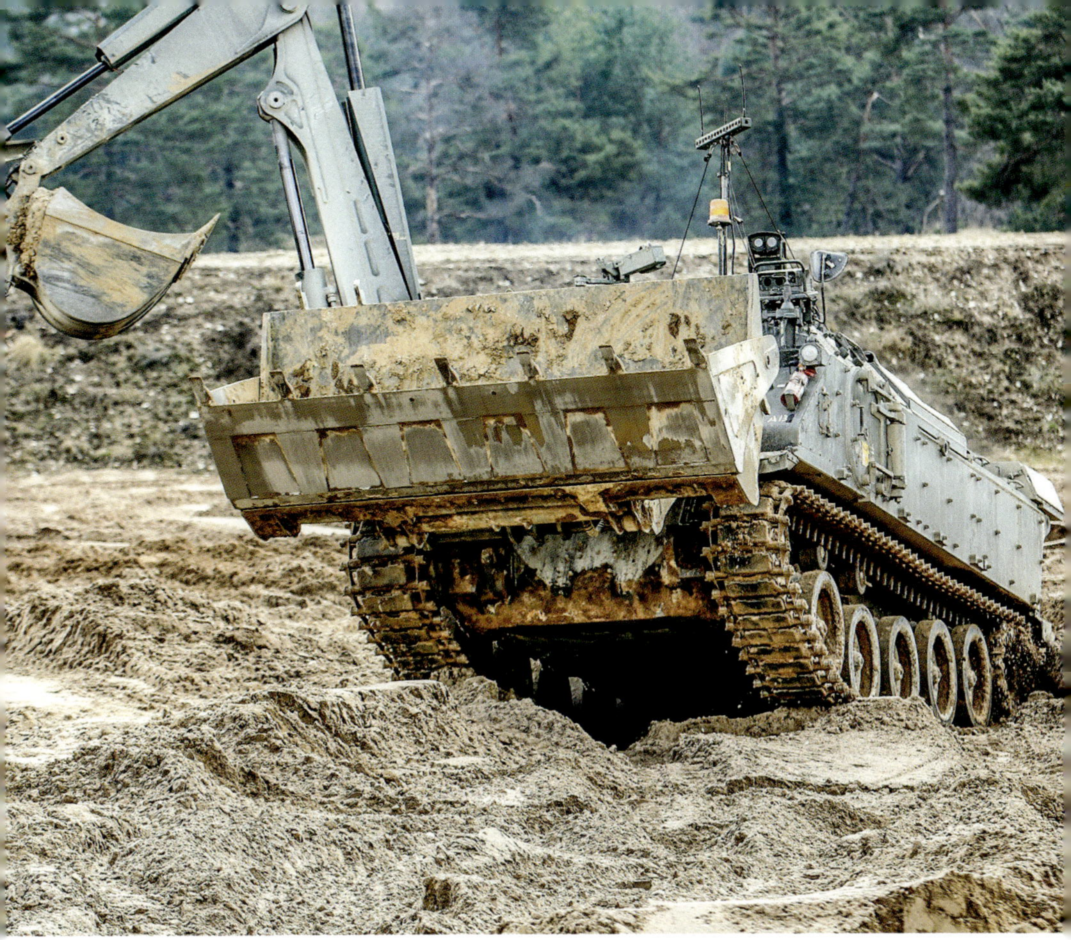

A robot clears a path before the soldiers arrive.

DID YOU KNOW?

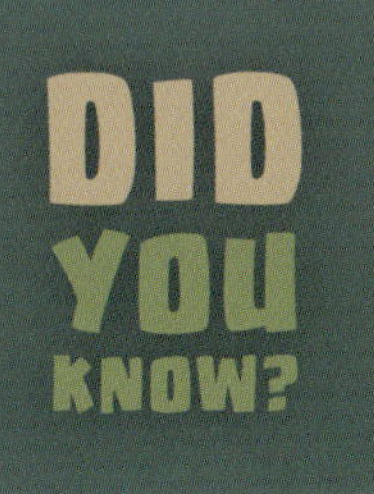

Before robots, soldiers had to clear paths themselves. This put them at risk of stepping on mines.

Military robots can do many tasks. Some protect soldiers or carry supplies. Some gather information. Other robots attack. Military robots range in size from tiny **drones** to huge tanks.

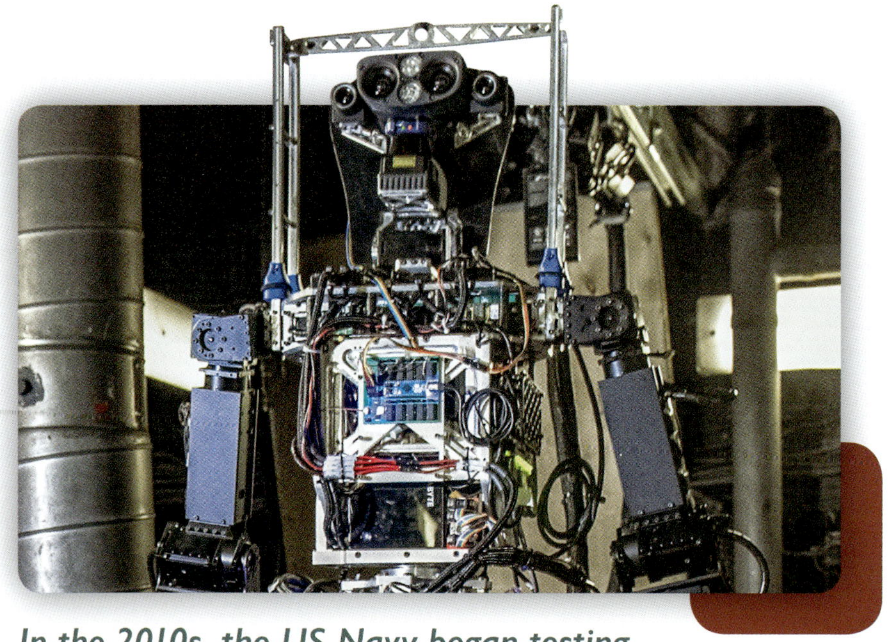

In the 2010s, the US Navy began testing a robot that can help sailors. The robot's human shape helps it get around the tight spaces of ships.

One drone can gather information in extreme weather. It can survive winds of up to 40 miles per hour.

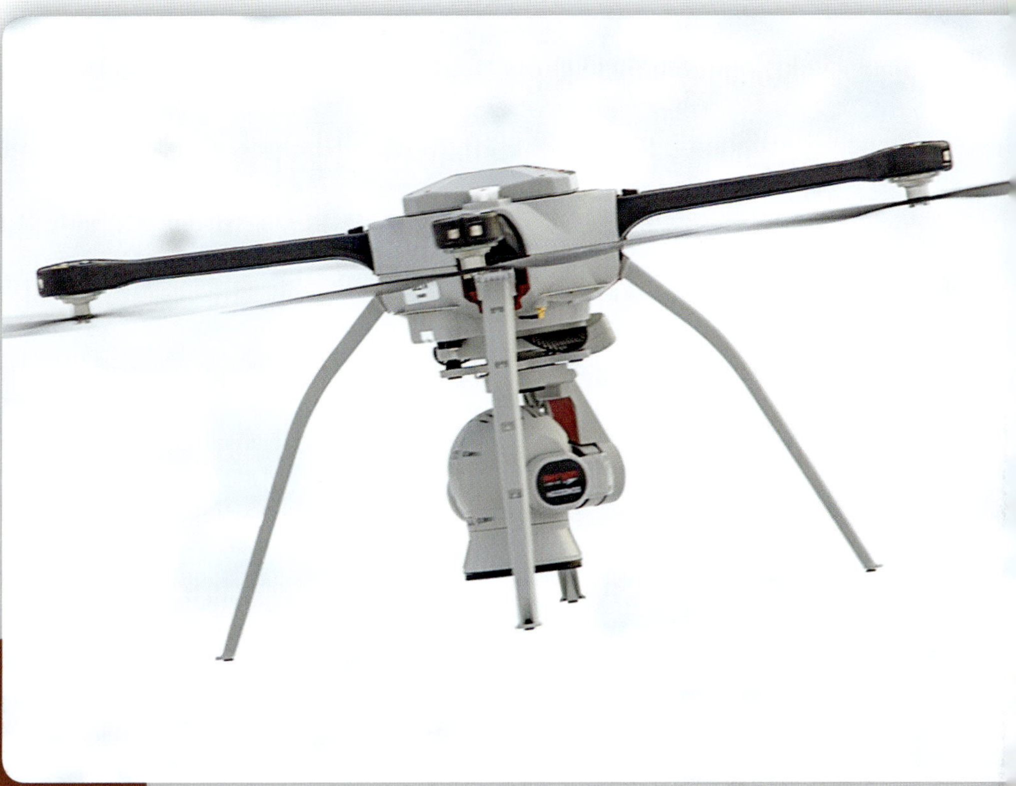

DID YOU KNOW?

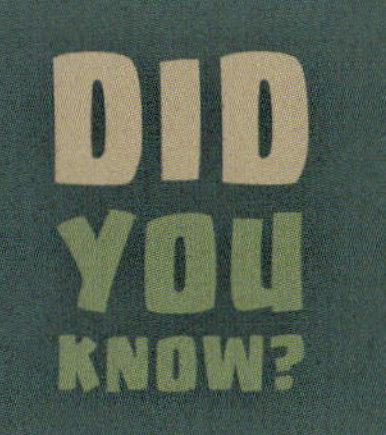

Some military robots help with fires. One robot throws an explosive to put out a fire.

CHAPTER 2

HISTORY OF MILITARY ROBOTS

Militaries began using robots in the 1900s. Many early robots were flying bombs. Others were **drones**. Soldiers used remotes and radios to steer them. But they were hard to control.

COMPLETE AN ACTIVITY HERE!

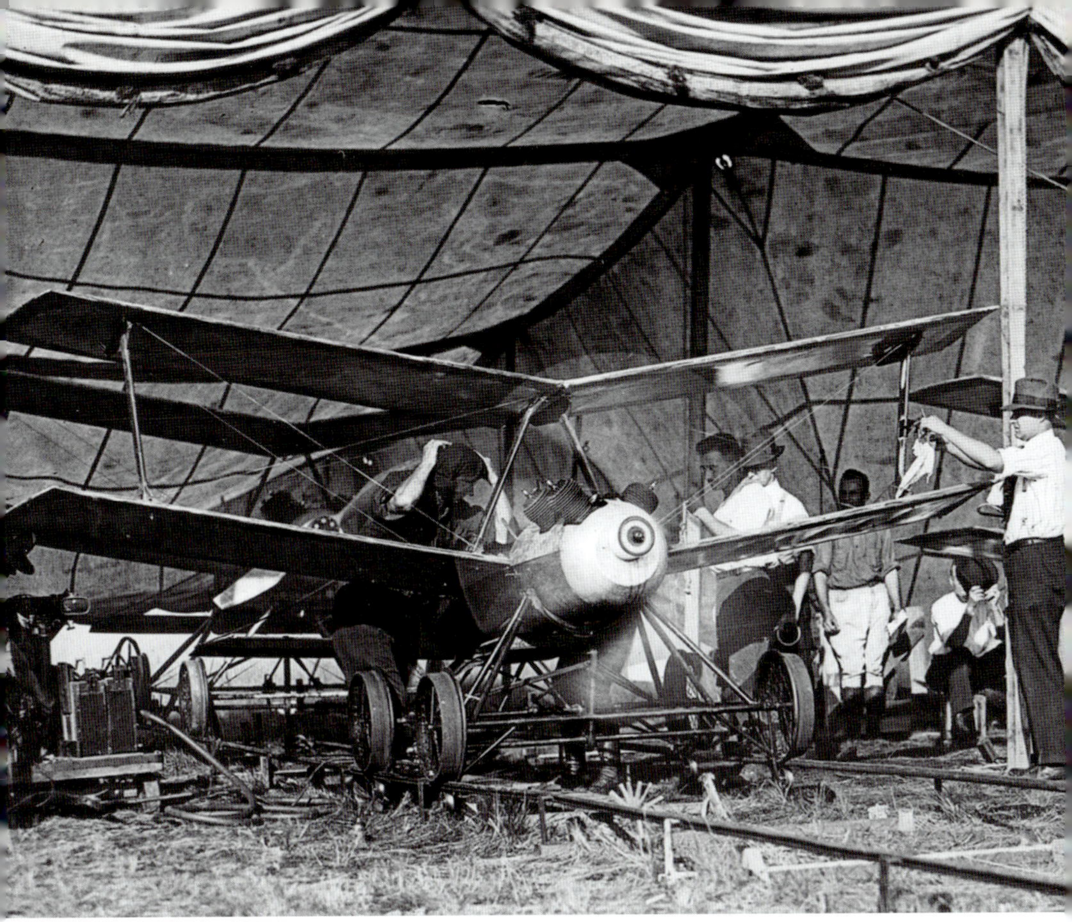

During World War I (1914–1918), one remote-controlled bomb was called Bug.

DID YOU KNOW?

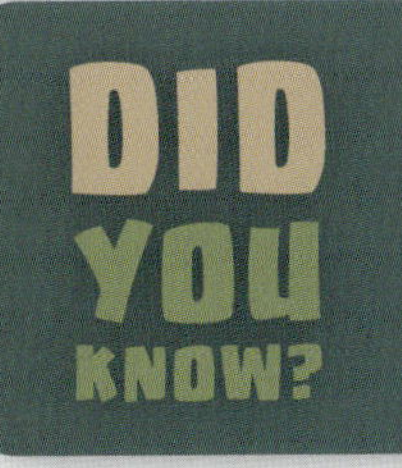

In the 1930s, pilots shot at drones for target practice.

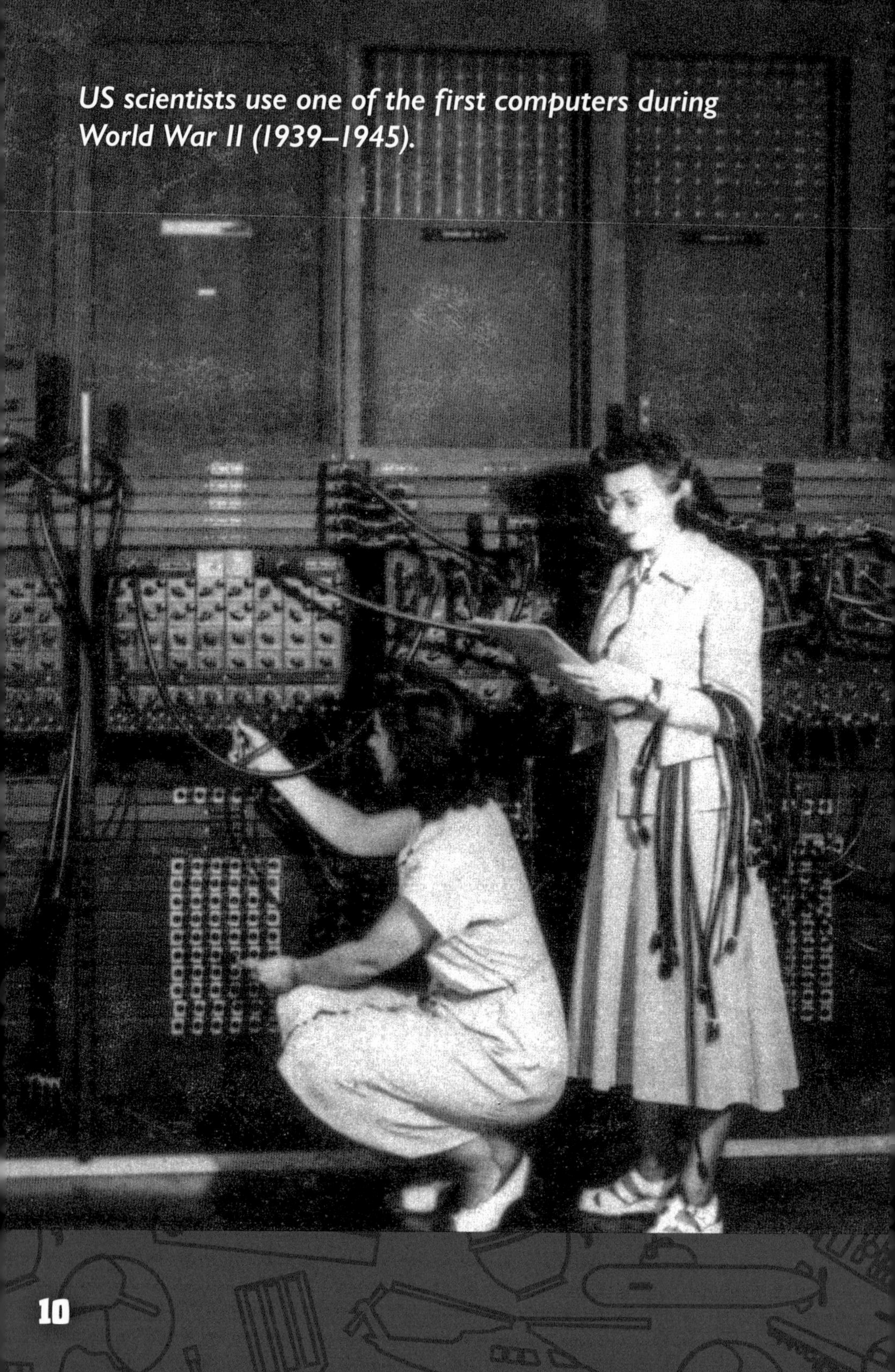

US scientists use one of the first computers during World War II (1939–1945).

Sensors and computers helped solve this problem. Sensors helped robots collect information. And computers organized it. By the 1990s, computer systems helped soldiers control robots.

In 1944, the US Navy flew one of the first drones that had a video camera.

The biggest change was **GPS**. This system shows an object's exact location. In 1995, people began using

The Predator drone was one of the first robots to use GPS.

GPS in robots. They could see exactly where each robot was. GPS made robots much easier to find and steer.

By the 2000s, **unmanned** aircraft had become common. Militaries began building other kinds of robots too. Some used weapons to fight. Others gathered information. Some detected harmful substances or bombs.

The first armed ground robot to see combat was called TALON.

An unmanned aircraft is launched into the air.

MILITARY ROBOTS TIMELINE

1914
Militaries use the first radio-controlled airplanes during World War I.

1943
Germany fires the first guided bombs during World War II.

1963
The US military uses drones to gather information during the Vietnam War (1955–1975).

1995
Militaries begin building robots that use GPS.

2007
The US military uses the first armed ground robots on the battlefield.

2014
US scientists build a drone the size of a penny.

CHAPTER 3

MILITARY ROBOTS TODAY

Robots are now part of many military missions. Armies often use **drones** to attack enemies. These drones carry **missiles**. They fire missiles at areas where the army believes enemies are.

LEARN MORE HERE!

Drones sometimes hurt innocent people. Some people think militaries should not use drones.

The soldiers who steer the drones can stay far from danger.

Some robots gather information. One, called PackBot, uses cameras to explore new areas. This robot can also safely get rid

DID YOU KNOW?

By 2018, the US Army had approximately 7,000 ground robots.

of bombs. Black Hornet looks like a tiny helicopter. Its cameras stream live video.

A soldier shows a Black Hornet in his hand.

A soldier uses a computer to control a robot.

Other robots help with fighting on the ground. SWORDS is an armed robot. It can carry a large gun or **grenade launcher**. It is controlled from a distance. SWORDS also has strong cameras.

SWORDS robot

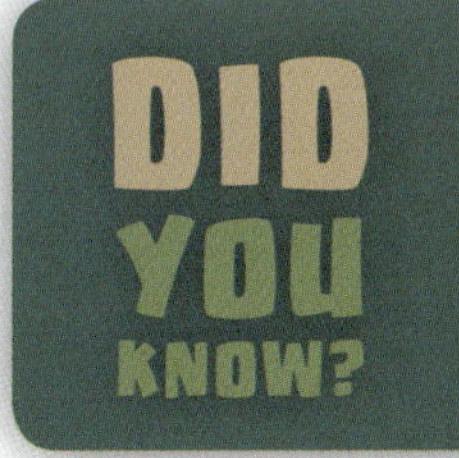

SWORDS's cameras can read a name tag more than three football fields away.

CHAPTER 4
SPECIAL MILITARY ROBOTS

MAARS is a newer robot based on SWORDS. MAARS is only three feet tall. But it has many weapons. Its sirens and lasers can distract people. And a **grenade launcher** flings explosives.

LEARN MORE HERE!

PARTS OF A MAARS ROBOT

The RoboBee acts like a tiny spy. It is the size of a bee. This little robot can fly and swim. Its cameras record sound and video. Another robot, the Cormorant,

can fly injured soldiers out of danger. It is **unmanned**. This robot may save pilots from the risks of rescue missions.

The Cormorant can fly 112 miles per hour.

Scientists are also working on new ideas for robots. Many new robots use artificial intelligence (AI). AI helps robots

In 2018, the British military tested a robot tank that used artificial intelligence.

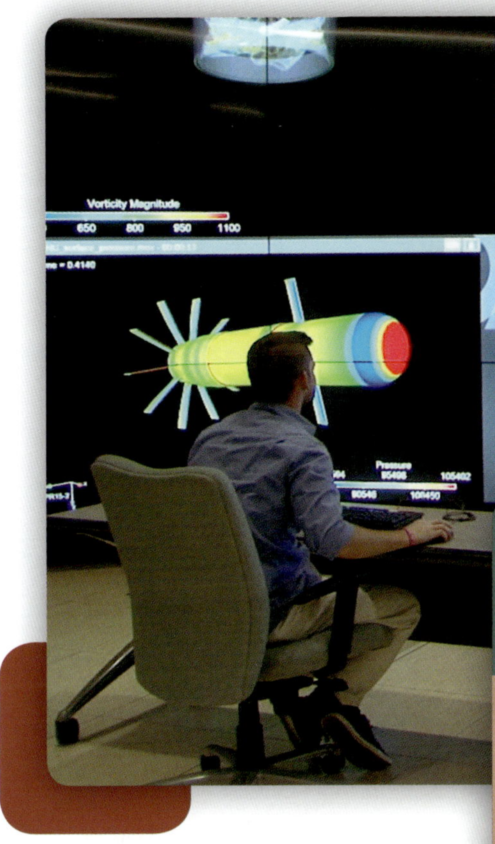

behave more like humans. The robots can learn and perform tasks on their own.

PROGRAMMING ROBOTS

Programming is a way of creating directions for computers. It tells a robot what to do. Robots first gather information about the world around them. Their programming then tells them how to respond. In the past, robots could only follow specific steps. But AI allows robots to learn or make choices.

MAKING CONNECTIONS

TEXT-TO-SELF

What kind of military robot would you like to try using? Why would you choose that one?

TEXT-TO-TEXT

Have you read books about other kinds of robots? How are they similar to and different from military robots?

TEXT-TO-WORLD

Military robots help soldiers. Can you think of other people that robots could help?

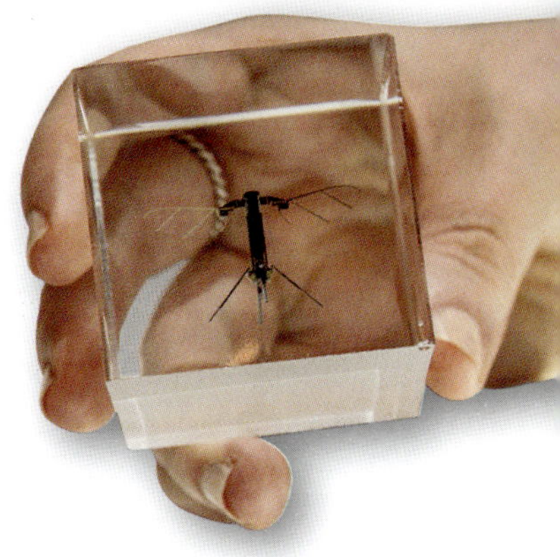

GLOSSARY

drone – an unmanned aircraft controlled by remote or by a computer.

GPS – short for "Global Positioning System," a system that uses information from satellites to determine an object's location on Earth.

grenade launcher – a tool that fires small, exploding weapons.

mine – an explosive weapon often put underground.

missile – a weapon that is thrown or fired.

unmanned – not controlled by someone touching the object.

INDEX

ONLINE RESOURCES

popbooksonline.com

Scan this code* and others like it while you read, or visit the website below to make this book pop!

popbooksonline.com/military-robots

*Scanning QR codes requires a web-enabled smart device with a QR code reader app and a camera.